HOW TO SOCIALIZE

REACTIVE DOGS

Transforming Reactive Behavior into Canine Confidence and Harmonious Companionship

WILLIAM C. KELLY

Copyright © 2024 by William C. Kelly

All rights reserved. Reproduction or transmission of any part of this book, whether by electronic or mechanical means, such as photocopying or recording, or through any information storage and retrieval system, is strictly prohibited without the explicit written consent of the author. The sole exception is for the incorporation of concise quotations in a review.

TABLE OF CONTENT

TABLE OF CONTENT……………………2-3
INTRODUCTION…………………………..4-8
CHAPTER ONE: The Importance and About the Book……………………………………....9-12
Understanding Reactive Behavior
Importance of Socialization
CHAPTER TWO: Recognizing Reactivity
Signs of Reactivity……………………….13-16
The Science of Canine Behavior
CHAPTER THREE: Building a Foundation……………………………....17-20
Establishing Trust with Your Dog
Basic Training Techniques
Gradual Exposure
Counterconditioning Techniques
CHAPTER FOUR: Socialization in Different Environments…………………………...21-24

Home Socialization

Training Tools and Aids

CHAPTER FIVE: Overcoming Challenges……………………………….25-28

Dealing with Setbacks

Success Stories: Real-life Examples of Socialization

CHAPTER SIX: Celebrating Progress………..………………………….29-31

Recognizing Achievements

Continuing the Journey

CONCLUSION…..…………………….32-36

APPRECIATION…..………………….37-40

INTRODUCTION

Welcome to "HOW TO SOCIALIZE REACTIVE DOGS," a comprehensive resource developed for dog owners seeking innovative strategies to confront and overcome the issues connected with canine reactivity. In the pages that follow, we start on a path of learning, training, and nurturing, trying to provide you with the information and resources required to steer your reactive dog towards a happy and socially adapted life.

In a world where each dog is as unique as its owner, the nuances of reactive behavior necessitate a sophisticated approach. Whether your canine partner displays minor discomfort or more severe reactive behaviors, this book is meant to meet you where you are in your journey and give concrete ideas for growth.

Unveiling the Canine Psyche: We start our investigation with deciphering the secrets of the canine mind. Understanding reactive behavior is not only about detecting symptoms but diving into the psychology that supports it. From behavioral triggers to genetic predispositions, we seek to provide you with knowledge that constitute the cornerstone for effective behavior management.

Building a Solid Foundation: Building trust is vital, and the early chapters assist you through creating a solid foundation of trust and communication with your dog. Learn key training strategies, establish a secure setting, and appreciate the necessity of progressive exposure to varied stimuli. These core concepts lay the framework for successful socialization and beneficial behavioral change.

Navigating Environments and Interactions: Socialization doesn't happen in solitude. We go into the subtleties of socializing reactive dogs in varied contexts, both at home and outside. From mastering interaction with other canines to choosing the correct training methods and assistance, you'll get practical insights to manage the intricacies of the real world.

Overcoming Challenges and Celebrating Success: Challenges are unavoidable, but this book doesn't simply identify setbacks, it includes solutions for overcoming them. Learn to manage aggressive behavior, seek expert advice when required, and celebrate the tiniest triumphs along the way. Real-life success stories give inspiration and practical direction, illustrating that the route towards socializing is both possible and gratifying.

As you explore the pages of "HOW TO SOCIALIZE REACTIVE DOGS," envisage a future where your canine friend thrives in social situations, demonstrating confidence and excellent conduct. Whether you're a rookie or an experienced dog owner, this book is your compass for navigating the rewarding route towards a well-socialized and fulfilled existence for you and your furry buddy.

Join me as I start on this transforming journey, one that honors development, perseverance, and the unbreakable relationship between you and your four-legged partner.

CHAPTER ONE: The Importance and About the Book

Welcome to "How to Socialize Reactive Dogs" a comprehensive resource devoted to helping dog owners manage the issues of reactive behavior in their canine friends. This book is aimed to give useful insights, practical tactics, and professional guidance on socializing dogs who demonstrate reactive tendencies. Whether you're a seasoned dog owner or a first-time caretaker, this book strives to equip you with the information and resources required to establish a healthy and well-adjusted relationship with your furry buddy.

In "How to Socialize Reactive Dogs," we dig into the nuances of canine behavior, concentrating on recognizing and resolving reactive tendencies in dogs. This book combines scientific insights with practical advice to give a comprehensive approach to socialization, adapting to the specific requirements of each dog and their owner. Packed with real-life examples, professional advice, and step-by-step training approaches, this book is your go-to resource for changing reactionary behavior into constructive social interactions.

Understanding Reactive Behavior

To successfully manage reactive behavior, it's vital to know the underlying elements leading to such behaviors in dogs. In this part, we study the indications and causes of reactive behavior, giving you the skills to analyze and understand your dog's behaviors. From environmental triggers to genetic predispositions, we dig into the science of canine behavior, presenting a comprehensive knowledge that lays the cornerstone for effective socialization.

Importance of Socialization

Socialization is a critical component of a dog's general well-being, impacting their behavior, adaptability, and overall contentment. This section stresses the vital importance that adequate socialization plays in building a well-adjusted and confident canine companion. We address the long-term advantages of socializing, not just in terms of behavior modification but also in developing pleasant connections with other dogs, humans, and the environment. By realizing the significance of socialization, you'll be ready to build the framework for a healthy cohabitation between you and your furry pet.

CHAPTER TWO: Recognizing Reactivity

Signs of Reactivity

Identifying reactive behavior in your dog is vital for successful intervention. Reactive behavior may show in different ways, such as increased barking, lunging, growling, or even violent behaviors. Understanding these signals is the first step in addressing and altering your dog's responses.

Common Triggers: Dogs display reactivity in reaction to certain stimuli. Common triggers include unexpected persons, other pets, loud sounds, or specific locations. Recognizing these triggers is vital for applying focused socialization tactics and building a specific training plan for your dog.

Assessing Your Dog's Reactivity Level: Each dog is unique, and their degree of sensitivity varies. This section leads you through analyzing your dog's reactivity level, including criteria like frequency, severity, and length of reactive behaviors. By carefully measuring your dog's behavior, you may adjust your approach to match their individual requirements.

The Science of Canine Behavior

Understanding Dog Psychology: To treat reactive behavior, it's vital to dig into the nuances of canine psychology. This section addresses the cognitive and emotional components of a dog's thinking, offering light on how they perceive and react to the environment around them. Understanding dog psychology creates the basis for successful behavior change.

Behavioral Triggers and Responses: Dogs display a variety of behaviors dependent on internal and external inputs. By researching behavioral triggers and reactions, we get insights into the motives driving reactive behavior. This understanding is helpful in devising proactive tactics to reduce unwanted behaviors and promote favorable ones.

The Role of Genetics in Reactivity: Genetics have a vital part in molding a dog's temperament and behavior. This section discusses how hereditary factors lead to reactive behaviors. Recognizing the role of genetics gives useful background for designing training techniques and recognizing the possible problems involved with altering specific habits.

CHAPTER THREE: Building a Foundation

Establishing Trust with Your Dog

Building a good foundation starts with creating trust between you and your animal friend. This section addresses the role of trust in creating a healthy relationship. We cover techniques for winning your dog's trust via continuous and good encounters, stressing the need of patience and understanding.

Basic Training Techniques

Foundational training is crucial for a well-behaved and socially adjusted dog. This part addresses essential training approaches that constitute the cornerstone of successful communication. From fundamental instructions to leash etiquette, adopting these approaches builds a foundation for effective behavior change and socialization.

Creating a Safe Environment: A safe and secure environment is vital for a dog's well-being and comfort. This section aids you in building an atmosphere that eliminates stress and encourages good behavior. We discuss factors such as adequate confinement, the necessity of mental stimulation, and the influence of the physical environment on your dog's general behavior.

Gradual Exposure

Introduction to Gradual Exposure: Gradual exposure is a cornerstone of socializing for reactive dogs. This portion teaches the notion of introducing your dog to triggers in a controlled and progressive way. We examine the reasoning behind progressive exposure and emphasize its efficacy in desensitizing dogs to stimuli that produce reactive behavior.

Desensitization Strategies: Desensitization is a methodical technique to decrease reactivity by progressively exposing your dog to trigger stimuli. This part presents realistic desensitization procedures, describing step-by-step approaches to accustom your dog to previously hard settings. Through these tactics, you'll learn to develop your dog's tolerance and confidence.

Counterconditioning Techniques

Counterconditioning includes modifying your dog's emotional reaction to trigger stimuli. This section goes into effective counterconditioning tactics that may be implemented into your training regimen. By matching pleasant experiences with previously aversive stimuli, you may modify your dog's connections and encourage positive behavior in many settings.

CHAPTER FOUR: Socialization in Different Environments

Home Socialization

Effective socialization starts at home, where your dog feels most comfortable. This section discusses ways for acclimating your dog to many parts of home life, including meeting new people, exposure to different noises, and navigating inside environments. Home socialization creates the framework for larger encounters in other situations.

Outdoor Socialization: Venturing into the big outdoors is an essential component of a well-socialized dog's existence. This part includes outdoor socializing, covering issues such as meeting strangers, traversing public settings, and reacting to environmental cues. Practical advice and activities are presented to help your dog adjust favorably to the varied outside world.

Interaction with Other Dogs: Positive encounters with other canines contribute greatly to a dog's social development. This section focuses on developing healthy dog-to-dog relationships, highlighting the significance of detecting canine body language and enabling regulated play. Insights on handling group settings and creating favorable encounters lead to a well-socialized and confident canine partner.

Training Tools and Aids

Collars and Harnesses: Selecting the correct collar or harness is vital for efficient training and control. This portion includes recommendations on picking suitable collars and harnesses depending on your dog's size, breed, and temperament. We examine the benefits of several solutions, from classic collars to no-pull harnesses, ensuring your training tools correspond with your dog's requirements.

Positive Reinforcement Tools: Positive reinforcement is an effective training approach that depends on rewarding desirable actions. This section presents a number of positive reinforcement strategies, including rewards, toys, and praise. Learn how to properly combine these techniques into your training routine, producing a good learning experience for your dog and increasing the link between you.

Professional Training Support: For more difficult behavioral concerns or if you prefer help from specialists, finding professional training support may be useful. This part explains the advantages of working with qualified dog trainers and behaviorists. We investigate how professional help may address unique obstacles, give individualized training strategies, and provide continuing advice for a successful socialization journey.

CHAPTER FIVE: Overcoming Challenges

Dealing with Setbacks

Socializing reactive dogs may offer hurdles, and failures are a normal part of the process. This section covers successful ways for handling setbacks, highlighting the value of patience and perseverance. Learn how to assess failures, change your strategy, and maintain a positive outlook to keep going on the road to effective socializing.

Handling Aggressive Behavior: Addressing aggressive behavior is a critical element of socializing reactive dogs. This part gives insights into identifying the core causes of hostility and developing techniques to control and improve this behavior. Practical ideas for guaranteeing safety, adopting behavior modification tactics, and encouraging alternate reactions are explored to create a safer and more pleasant atmosphere.

Seeking Professional Guidance: For difficult or chronic behavioral disorders, getting professional advice is a useful step. This section discusses the advantages of consulting with qualified dog trainers, behaviorists, or veterinarians. Discover how professional experience can give specific solutions, individualized training programs, and continuous assistance, boosting the probability of successful socialization for your dog.

Success Stories: Real-life Examples of Socialization

In this exciting chapter, we highlight success tales of dogs who have overcome reactivity and matured into well-socialized companions. Each tale illustrates the particular problems experienced by the dog and their owner, documenting the techniques taken, the milestones accomplished, and the transforming effect of successful socialization. These real-life examples serve as encouragement and assistance for readers on their own socialization journey.

Building a Supportive Community: Socialization is not a solo path; it thrives on group support. This section highlights the necessity of developing a network of like-minded persons who understand the difficulties and achievements of socializing reactive dogs. From online forums to local training groups, learn how a supportive community may give encouragement, assistance, and a feeling of camaraderie throughout the socializing process.

CHAPTER SIX: Celebrating Progress

Recognizing Achievements

As you commence on the path of socializing your reactive dog, it's crucial to appreciate every milestone and success. This part invites you to reflect on the progress accomplished, no matter how modest. Recognizing and applauding your dog's triumphs promotes morale, encourages good behaviors, and enhances the link between you and your furry friend.

Positive Reinforcement for Both You and Your Dog: Celebrating success is not just about honoring your dog's achievements but also appreciating your efforts as a loyal owner. This section discusses the notion of positive reinforcement for both you and your dog. By providing a happy and encouraging atmosphere, you create a mutually beneficial experience that fosters continuous dedication to the socialization process.

Continuing the Journey

Lifelong Socialization: Socialization is a continual process that goes beyond the early phases of training. This portion highlights the necessity of lifelong socialization to preserve and develop your dog's beneficial habits. Explore ways for integrating sociability into your regular routine, adjusting to new situations, and continuously rewarding good interactions.

Expanding Experiences: As your dog develops in their socialization journey, consider increasing their experiences to guarantee well-rounded growth. This section gives information on introducing new challenges, surroundings, and activities to keep your dog interested and adaptable. A varied variety of experiences adds to a confident and socially skilled canine companion.

Strengthening the Bond: Continuing the socialization process is a chance to build the link between you and your dog. Explore engaging activities, games, and shared experiences that improve the bond between you two. This section illustrates the beneficial influence of an enhanced connection on your dog's general well-being and behavioral development.

CONCLUSION

As we finish the last pages of "HOW TO SOCIALIZE REACTIVE DOGS," I hope this thorough book has been a light of information and inspiration on your road towards promoting good behavior in your canine partner. Socializing reactive dogs is a dynamic process, and your devotion to this attempt indicates not just your love for your furry buddy but also your determination to establish a peaceful and fulfilling existence for them.

Celebrating Progress: Take time to reflect on the leaps you and your dog have made together. Celebrate the development, both great and little, since each success is a testimonial to the trust, patience, and understanding you've established.

Recognizing these achievements deepens the good link between you and your canine friend and prepares the road for continuing success.

A Lifelong Adventure: Socialization is not a goal; it's a lifelong adventure. As you continue to explore new surroundings, negotiate various relationships, and adjust to altering situations, remember that the trip itself is as important as the goal. Embrace the continual process of learning and developing with your dog, and cherish the shared experiences that enhance your special relationship.

Developing a Supportive Community: Throughout this book, we've underlined the significance of developing a supportive community. Whether it's via online forums, local training organizations, or the shared companionship of other dog owners, a supporting network may give encouragement, assistance, and a feeling of understanding. As you go ahead, consider how you might contribute to and benefit from this network of like-minded people.

A Positive Future: Envision a future where your once-reactive dog navigates the world with confidence and delight. By implementing the concepts of understanding, trust-building, and positive reinforcement, you've created the framework for a great future. Your attention to socialization not only helps your dog's quality of life but also extends your own experience as a responsible and loving owner.

In summary, I offer my deepest appreciation for allowing "HOW TO SOCIALIZE REACTIVE DOGS" to be a part of your journey. May the wisdom inside these pages continue to lead you and your furry buddy towards a future filled with shared experiences, mutual understanding, and the lasting delight that comes from a well-socialized and pleased canine companion.

Remember, the tale doesn't stop here, it's a constant narrative of development, connection, and the endless possibilities that open when the love between a person and their dog is at the core of the trip.

Wishing you and your dear partner a lifetime of pleasure, shared adventures, and the contentment that comes from a life well-lived together.

APPRECIATION

Dear Reader,

It is with great joy and a strong feeling of dedication that I bring to you "HOW TO SOCIALIZE REACTIVE DOGS." As a passionate dog lover, trainer, and champion for the well-being of our furry companions, this project is a result of years of expertise, enthusiasm, and a shared journey with numerous dogs and their loving owners.

The Idea Behind the Guide: The idea for this guide derives from a thorough awareness of the specific issues experienced by owners of reactive dogs. Every dog is an individual, and each one deserves the chance to grow in a supportive and stimulating home. It is my aim that the ideas, tactics, and real-life experiences presented in these pages can empower you on your way to changing reactionary behavior into good social interactions.

A Holistic Approach to Socialization: Socializing reactive dogs is a multidimensional activity that goes beyond training approaches. It involves a profound awareness for the subtleties of canine behavior, the patience to negotiate hurdles, and the dedication to building a solid foundation founded on trust and understanding. This book seeks to give you with a complete toolset, empowering you to handle the individual requirements of your dog and building a relationship that transcends barriers.

Celebrating Your Role as a Dog Owner: Being a responsible dog owner is a journey of ongoing learning, adaptability, and love. In these pages, I salute your commitment to the well-being of your four-legged buddy. Your devotion to learning, training, and socializing your dog indicates not just your passion for animals but also your desire to provide a happy and rewarding life for them.

The Power of Community: As you traverse the difficulties of socialization, realize that you are not alone. Whether you find assistance in local training groups, internet forums, or the shared experiences of other dog owners, the power of community is important. Your path is unique, but the combined knowledge of a supportive network may give encouragement and insight when you need it most.

Wishing You a Fulfilling Journey:

As you flip the pages of this book, I wish you and your canine partner a journey filled with pleasure, development, and the profound satisfaction that comes from watching good improvements. May the link between you and your dog continue to deepen, and may the ideals discussed here contribute to a future where both you and your furry buddy flourish.

Thank you for entrusting me with a part in your journey as a dog owner. May "HOW TO SOCIALIZE REACTIVE DOGS" be a beneficial companion on your way to a joyful and satisfying existence with your cherished canine friend.

Warm regards.

Job websites, freelancer job websites, interview techniques and how to get hired tips

INTRODUCTION 1

THE ROLE OF JOB WEBSITES IN MODERN JOB HUNTING.......................2

NAVIGATING THE INTERFACE: HOW JOB WEBSITES WORK...................3

CRAFTING AN IMPRESSIVE PROFILE..4

THE POWER OF NOTIFICATIONS 5

BENEFITS FOR EMPLOYERS 6

TRANSITIONING TO A NEW CAREER PATH..7

SUPPORTING THE JOB MARKET 8

UPWORK: THE GATEWAY TO GLOBAL OPPORTUNITIES 9

DESIGNHILL: WHERE CREATIVITY MEETS COMMERCE 10

TOPTAL: ELEVATING ELITE FREELANCERS 11

LINKEDIN AND LINKEDIN SERVICES MARKETPLACE: THE POWER OF PROFESSIONAL NETWORKING 12

WE WORK REMOTELY: EMBRACING REMOTE WORK CULTURE 13

BEHANCE: SHOWCASING YOUR CREATIVE PORTFOLIO 23

SIMPLYHIRED: SIMPLICITY IN JOB HUNTING.......................................25

DRIBBBLE: THE DESIGNER'S PARADISE..26

FIVERR: YOUR MICRO-ENTREPRENEURSHIP JOURNEY 28

PEOPLEPERHOUR: HOURLY PROJECTS, LIFETIME OPPORTUNITIES 29

GURU: WHERE FREELANCERS AND EMPLOYERS UNITE 20

FREELANCER.COM: YOUR ONE-STOP FREELANCING HUB 21

DESIGNCROWD: CROWD-POWERED DESIGN EXCELLENCE............22
WELLFOUND: NAVIGATING NEW OPPORTUNITIES.................23
99DESIGNS: DESIGNING YOUR SUCCESS STORY 37
WORKING NOT WORKING: WHERE CREATIVE MINDS THRIVE 38
WEBFLOW EXPERTS: MASTERING THE WEBFLOW PLATFORM.....40
YUNOJUNO: SIMPLIFYING FREELANCER-CLIENT CONNECTIONS 41
AUTHENTIC JOBS: FINDING YOUR DREAM FREELANCE JOB 43
TASKRABBIT: EMPOWERING LOCALIZED FREELANCING 44
FLEXJOBS: YOUR SOURCE FOR REMOTE AND FLEXIBLE JOBS 30
SOLIDGIGS: CURATED FREELANCE SUCCESS 31
CONCLUSION..32

1. Indeed (www.indeed.co.uk): 33
2. Reed (www.reed.co.uk): 34
3. LinkedIn (www.linkedin.com/jobs):............................35
4. Totaljobs (www.totaljobs.com): 350
5. Monster (www.monster.co.uk): 37
6. CV-Library (www.cv-library.co.uk): 38
7. Jobsite (www.jobsite.co.uk): 39
8. Guardian Jobs (jobs.theguardian.com):.....................40
9. Gumtree (www.gumtree.com): 41
10. CWJobs (www.cwjobs.co.uk): 42
11. Glassdoor (www.glassdoor.co.uk): 43
12. Jobs.ac.uk (www.jobs.ac.uk): 44
13. Workinstartups.com (www.workinstartups.com): 45
14. Joblift (www.joblift.co.uk):....................................46

15. Fish4Jobs (www.fish4.co.uk): 47
16. JobsNHS (www.jobs.nhs.uk): 453
17. eFinancialCareers (www.efinancialcareers.co.uk): 49
18. UK Staff Search (www.ukstaffsearch.com): 50
19. JobisJob (www.jobisjob.co.uk): 51
20. CharityJob (www.charityjob.co.uk): 52
21. The Dots (the-dots.com/jobs): 53
22. Milkround (www.milkround.com): 54
23. BritishJobs (www.britishjobs.net): 55
24. RetailChoice (www.retailchoice.com): 56
25. Adzuna (www.adzuna.co.uk): 57
26. StudentJob UK (www.studentjob.co.uk): 58
27. TechNation Jobs (jobs.thetechnation.com): 59
28. Law Society Gazette (jobs.lawgazette.co.uk): 60
29. Engineering Jobs (www.engineeringjobs.co.uk): 61
30. Hired (hired.com): 62
31. Guardian Careers (jobs.theguardian.com/careers): 63
32. Construction Jobs (www.constructionjobs.co.uk): 64
33. Charity People (www.charitypeople.co.uk): 65
34. SecsInTheCity (www.secsinthecity.co.uk): 66
35. Media Match (www.media-match.com): 66
36. All Cruise Jobs (www.allcruisejobs.com): 67
37. Graduate Jobs (www.graduate-jobs.com): 68
38. Aviation Job Search (www.aviationjobsearch.com): 69
39. Creativepool (www.creativepool.com): 70

40. Jooble (www.jooble.org.uk): 71
41. Jobserve (www.jobserve.com): 72
42. AllAboutCareers (www.allaboutcareers.com): 73
43. Jobsgopublic (www.jobsgopublic.com): 74
44. IT Jobs Watch (www.itjobswatch.co.uk): 75
45. RetailChoice (www.retailchoice.com): 76
46. Caterer.com (www.caterer.com): 77
47. JustEngineers (www.justengineers.net): 78
48. Public Sector Jobs (www.publicsectorjobs.org.uk): 79
49. Travelweekly Jobs (www.travelweekly.co.uk/jobs): 80
50. Bubble Jobs (www.bubble-jobs.co.uk): 81

CHAPTER 1: THE ART OF IMPRESSIVE RESUMES 82
CHAPTER 2: CRAFTING A COMPELLING PERSONAL BRAND 83
CHAPTER 3: PREPARING FOR INTERVIEW EXCELLENCE 84
CHAPTER 4: MASTERING NONVERBAL COMMUNICATION 85
CHAPTER 5: NAVIGATING CHALLENGING SCENARIOS 86
CHAPTER 6: THE FOLLOW-UP: SEALING THE DEAL 867
CONCLUSION: YOUR PATH TO SUCCESS 90

    CHAPTER 1: PAVING YOUR PATH TO INTERNATIONAL EMPLOYMENT 91
    RESEARCHING GLOBAL INDUSTRIES AND TRENDS 92
CHAPTER 2: CRAFTING AN INTERNATIONAL-FRIENDLY RESUME 93
    SHOWCASING LANGUAGE PROFICIENCY 94
    CHAPTER 3: NAVIGATING JOB SEARCH PLATFORMS AND RESOURCES 95

www.ingramcontent.com/pod-product-compliance
Lightning Source LLC
Chambersburg PA
CBHW050249230526
45470CB00005B/2181